Machine Learning for Beginners

Easy Guide Book

Table of Contents

Introduction

Computers can be used for creation of systems which act like human beings. This is achieved by use of the concept of machine learning. Computers are able to perceive the environment and respond to the stimuli accordingly. In machine learning, you can take advantage of this to create a system which acts like a human. A good example is a self-driven car. Such a car will be able to sense objects and move away from them to avoid collisions. Machine learning can also be applied in business. An example of this is in the association of items which are purchased together. This shows machine learning can help businesses grow. This book explores machine learning in detail. Enjoy reading!

Chapter 1- What is Machine Learning?

Machine learning refers to the science of getting computers act without being explicitly programmed. This science has given us practical speech recognition, self-driven cars, a greater understanding of the human genome, and an effective web search. Machine learning is so wide that you use it on a daily basis without even noticing. Researchers have agreed that machine learning will help us progress to the human-level of artificial intelligence.

Machine learning is just a field of artificial intelligence (AI), which involves the study of how the computer agents may improve their cognition, perception, and action from their experience. Machines are expected to improve from their knowledge, data, experience, and interaction.

Machine learning can have a significant effect on many applications which rely on various types of data which have been recorded in a computer such as scientific data, health data, location data, financial data, energy data, weather data, and other types of data. This makes machine learning very crucial to both the current and the future computer applications.

Chapter 2- Machine Learning Algorithms

Machine learning algorithms can be classified into two broad categories, which include **supervised** and the **unsupervised** learning algorithms. The difference between these two categories of algorithms is the way in which training examples are fed into the algorithms. Let us discuss these.

Supervised Learning

In this type of learning, you can view the machine learning algorithm as a process which will transform input provided to it into the output which is desired. In this case, the machine learning process must learn the best way of transforming each possible input into the desired output, meaning that each example has the desired input as well as the desired output.

Consider the situation in which you have an artificial chess player. The input to this would be a chess board state, and the output to this would be best possible movement.

There are two subtypes of supervised learning which are determined by the output type. These include classification and regression.

Classification

This is the case in which the output value belongs to a discrete and finite set.
To make this easy for your understanding, in this type of algorithm, you have a set of classes and a set of data, and each data should be classified or mapped into a particular class. The output, which can be seen as classes, is made up of a set of finite options. A good example of this is a case in which you need to classify a set of statements as positive, negative, or neutral. These three are the classes or options, and the statements are the inputs which are to be mapped to the classes.

Regression

A regression problem occurs when the output value is a continuous number such as a probability. A good example of such a problem is when you have some individuals with debts, and you want to determine the probability that an individual will pay his or her debts. In this case, the probability that an individual will pay his or her debt will be between 0 and 1, and this can be classified as a regression problem. Note that supervised learning is the most widely used type in learning algorithms. However, it has a weakness in that for each training example, we have to provide the correct output, which in most cases might be very expensive.

Unsupervised Learning

This is another category of machine learning algorithms. In these algorithms, training examples have to form the input to an algorithm, but not the desired output.

The goal in this case is to determine the hidden relations and structure between our training examples. Clustering algorithms are a good example of unsupervised learning algorithms, in which we need to determine similar groups or instances of instances or clusters.

Reinforced Learning

In this type of learning, the algorithm is given an observation of the world, and it then learns a policy regarding how to act. Each action should have an effect to the environment, and the environment will give back some feedback which will help the algorithm to learn. The algorithm is expected to choose an action depending on the data point, and it later learns whether the decision was good or not.

Those are the three classes or categories of machine learning algorithms. Each machine learning algorithm can be classified or mapped to any of the above classes. We can now discuss some of the available machine learning algorithms.

Common Machine Learning Algorithms

Let us discuss some of the common machine learning algorithms:

Naive Bayes Classifier Algorithm

A classifier refers to a function which allocates an element value of a population from any of the categories which are available. When it comes to a document, a web page, an email and other lengthy text notes, it becomes hard for us to manually classify them. The Naïve Bayes algorithm comes to help us in this case. A good example of application of this algorithm is in Spam Filtering. The Spam Filter works by assigning the label "Not Spam" or "Spam" to any email.

The Naïve Bayes classifier is among the popular learning methods which have been grouped by similarities, and it works based on the Bayes Theorem of probability in building machine learning models most probably for document classification and disease prediction.

The algorithm should be used under the following circumstances:

1. When you are provided with a large or moderate training data set.

2. If there are several attributes for the instances.

3. When the attributes describing the instances are expected to be conditionally independent.

This is the algorithm used in Sentiment analysis, such as Facebook, to analyze the status updates which express either positive or negative feelings.

It is also applied in PageRank, whereby certain pages are marked to be important while in databases. The algorithm is also widely applied in the classification of new articles as Sports, Technology, Politics, or other categories.

K-Means Clustering Algorithm

This is an unsupervised learning algorithm used in machine learning for cluster analysis. K-Means is an iterative and nondeterministic method. The algorithm normally works on a given data set through a defined number of clusters, denoted as k. The output in this algorithm is the k clusters with the input data already partitioned into necessary clusters.

A good example of an application is the Wikipedia search results. Suppose you search for the word "Jaguar" on the Wikipedia website. All the pages having the word "Jaguar" as an animal as well as those having "Jaguar" as a version of the Mac OS will be displayed.

You will also get all the pages with "Jaguar" as a car. In such a case, the K-Means algorithm can be applied to group together all the pages which are referring to the same concepts. In this case, there will be three clusters, one about the animal, another one about the car, and the last one about the Mac OS version. Each web page in the search results will be mapped to the right cluster.

The main application of the K-Means algorithm is in the search engines such as Yahoo and Google for clustering the web pages based on similarity and determining the "relevance rate" of the search results. This is good, as it helps the search engines to reduce the computational time for the users.

Apriori Machine Learning Algorithm

This is an unsupervised learning algorithm which works by generating association rules from a given data set. The association rules state that if item X occurs, then an item B occurs in some certain probability. Most association rules which are generated are usually in the form of the IF_THEN form. Example, if people buy a phone, then they also buy a case for protection. For your algorithm to arrive at such a conclusion, it has to observe the number of people who have bought the case while buying a phone. A ratio should then be derived from this, for example, out of 50 people who bought a phone, 40 of them also bought a case.

The Apriori algorithm is highly applied in Market Basket Analysis. This is a technique used in most e-commerce websites such as Amazon to determine items which are purchased together. A good example is when you buy a book about a computer programming language such as Python on Amazon. You will also get an email from Amazon recommending that you buy a related book, such as a Linux book.

The algorithm is also widely applied in Auto-Completion applications. You must have seen this as you search on your browser. When you type a particular word, other words which are usually together with that word will be recommended, and you will see them.

Linear Regression Machine Learning Algorithm

This algorithm is used to show the relationship between two variables and how a change in one of the variables will affect the other. Simply, it shows how the dependent variable is affected after a change in the independent variable.

This algorithm is widely applied in the business of forecasting sales based on trends. After the company has observed a steady increase in sales during some months, the algorithm will be applied to help predict the sales for the coming months.

Decision Tree Machine Learning Algorithm

A decision tree refers to a graphical representation which uses the branching methodology to determine all the possible outcomes for a decision depending on a number of conditions. In the decision tree, the internal nodes usually represent a test on an attribute, and each branch in the tree is a representation of the outcome from a test, while the leaf node is a representation of a class label, that is, a decision which has been reached after a computation of all the possible outcomes. With decision tree machine learning algorithms, we are able to make decisions under conditions of uncertainty. Due to their visual representation, communication is improved. With a decision tree, it is also possible for you to determine what the outcome would have been if you had chosen a different decision.

They are best suited in circumstances where there are errors in the training data. A decision tree can also help you determine the value for a missing data by observing the data contained in the other tables. Decision tree machine learning algorithms are highly applied in banks for classification of loan applicants based on the chances of defaulting payment.

Chapter 3- Perceptron's, Logistic Regression, and SVMs

Perceptrons

AI (artificial intelligence) and machine learning are geared towards doing what humans are able to do and surpass them. This calls for copying of what makes the humans to be so great, which is the brain.

The brain is made up of billions of neurons which are interconnected, and these keep on sending signals from one neuron to another. These neurons work together so as to help us recognize patterns, recognize faces, and think.

The neuron operates from a very simple principle. In case the inputs have matched a particular criteria, then it will fire, and the signals will be sent to the rest of the neurons which are connected to it. In machine learning, we try to use the computer and simulate the brain, so we only have to copy the basic idea of how a brain works.

The perceptron emulates the neurons' functions in the machine learning algorithms. It forms one of the oldest machine learning algorithms which are available. It was first applied in rudimentary image recognition in 1957. Currently, the perceptron forms the basis for most neural networks which are being used in the world.

The perceptron works by taking many inputs and giving the output. A good example of this is a perceptron which takes in temperatures and gives an answer to the quest, "Will I wear my sweater today?" If you have a threshold of 65 degrees, and the input temperature happens to be below this, the output should be 1 (yes), while if the temperature is above this, it will output 0 (No).

However, it is possible for us to consider parameters other than temperature when we need to decide whether to wear a sweater or not. Our biological neurons are capable of receiving more than one electrical impulse, and the same can be applied to the perceptron. Each input will also be given a weight. Suppose we use the wind speed, temperature, and the population of people taking a shower in a town such as Lagos as the inputs, different weights would be assigned to the different inputs.

When temperatures are low, there is a high probability that we will wear a sweater, so the value of the temperature in this case would be negative. When the wind speeds are high, we will need to wear a sweater, so this will mostly have a positive weight. In the case of a population of people taking a shower in Lagos, the weight should be zero. Our assumption is that Lagos is a hot town.

Suppose you are talking about the population of individuals living in Canada. A person living in such a place might have been used to low temperatures. This means that their threshold for wearing a sweater will be at a lower temperature compared to someone living in a town in Australia.
This will be best expressed by use of a bias which will help us show the threshold for both the Canadian and the Australian examples. The bias can be seen as a measure of the difficulty of the perceptron saying either "yes" or "no."

Logistic Regression

The perceptron has some weaknesses when it comes to learning. It views life as either white or black, which is not the case. For instance, you will not wear the sweater immediately when the temperature has dropped below a certain threshold. The probability of wearing a sweater at a certain temperature can be expressed as a percentage, for example, when the temperature is at 35 degrees, an individual will have a 90% chance of wearing a sweater.

Modern life is full of complexities. With logistic regression, we can model these complexities so as to fit in a probability. A logistic curve is fed to the data. The gradient descent is used for choosing the best parameters for a model. Our aim is to get the parameters which will help to minimize the cost function.

The logistic regression and the perceptron algorithm are closely related. The logistic regression can be seen as a perceptron algorithm working on steroids. The logistic model is capable of predicting probabilities, while a perceptron is only capable of predicting either yes or no.

Support Vector Machines

The SVMs, or Support Vector Machines, are machine learning algorithms which have become very popular in recent days. Let us discuss some of the ideas of this algorithm.

Margin

In a classification algorithm, data is separated by drawing a line, commonly known as the decision boundary. This is shown below:

Once the decision boundary has been drawn, it becomes easy for us to determine the margin for every data point. The margin for each data point refers to the distance from it to the decision boundary. The margin helps us determine our level of confidence in the classification.

If the value is located far from the decision boundary and has a large margin, then we will be a bit more confident with our prediction. However, for a value which is located close to the decision boundary and has a low margin, we will not be very sure with our classification.

Now that you know what margins are, it is good for you to know what support vectors are. A support vector refers to a vector from a data point having a smallest margin from a decision boundary. To make it simple, it refers to the vector between the data point which is closest to decision boundary and the boundary itself. Note that this vector is usually perpendicular to the decision boundary, since the smallest distance existing between the point and the line is a perpendicular line.

The support vector machine is aimed at classifying data by the drawing of a decision boundary in such a way that it will maximize support vectors. If we maximize the support vectors, we will also have maximized the margins in our data set, and the decision boundary will be located very far from our data points.

Linear Reparability

For those who have played around with simulations, you will finally conclude that the algorithm fails at some point.

This will only happen when your data points are not linearly separable. In cases where you are unable to draw a straight line to separate your data, you will have data which is linearly inseparable. Due to the inability to draw a straight line separating the data, the SVM algorithm will fail.

We then have to come up with a mechanism which will help us deal with the linear inseparable data. We are able to reformulate the optimization problem.

Initially, our aim was to have each data point located as far from the decision boundary as much as possible. In our case now, we want to allow the data point to stray to the wrong side of our boundary, but we will have to add some cost to our process.

This is a common occurrence in machine learning, and it is referred to as "regularization." It is a good idea for allowing us to have a flexible model when we are classifying our data. When we violate this, the decision boundary can be too high or too low, and we use a "regularization parameter" to control it.

Mathematically, regularization is implemented by the addition of a term to the cost function.

Chapter 4- Neural Networks for Machine Learning

Neural networks are among the recent developments in machine learning. They can be applied in solving the majority of the problems in today's world. Do you need a self-driven car? Neural networks are the best approach to help you make one. You can also apply the concept of neural networks to fly a helicopter. Neural networks are a promising algorithm.

Neural networks were developed based on the concept of how the human brain works. The biological brain can be viewed as the most powerful computer in existence. In the year 2014, researchers in Japan tried to simulate a single second of the human brain by use of a supercomputer. This took a total of 40 minutes and 9.9 million watts. This shows that the human brain is more powerful than a supercomputer, with the former running on only 20 watts.

The biological neurons found in the human brain work by receiving signals from other neurons, and then they forward these signals to the other neurons which they have been connected to. The neuron usually has a certain threshold, and before this threshold is attained, the neuron cannot fire or propagate the impulse. The neuron continues to receive signals from the other connected neurons, and once the threshold is attained, it will propagate or fire the impulse to the other neurons connected to it.

This continues on and on. The threshold for a particular neuron is different from that of other neurons, and it is determined by the chemical composition of the neuron. The brain of a human being has billions of these neurons which are interconnected and communication between of all these neurons forms the basis for thought, consciousness, and the McChicken cravings.

History of Neural Networks

In the 1900's, researchers decided to create a mathematical model to simulate the function of the human brain. The first model to be created was a simulation of a single neuron, and it simulated the inputs, outputs, and the threshold for the neuron. The output from the neuron was again fed into more artificial neurons, and an artificial neural network was formed.

However, the researchers experienced a challenge regarding the model. Though they had created a model of the human brain, they lacked a way of teaching it. There was no way of getting a logical output from the artificial brain. The researchers saw a need to come up with an algorithm which would help the artificial brain to learn.

In the 1980's, a learning algorithm for the artificial brain was developed, and this was named the "backpropagation algorithm." This algorithm allows artificial neural networks to be trained so as to do amazing things, such as the development of self-driven cars.

How Neural Networks Work

If you remember what we discussed about the perceptron, then it will be easy for you to understand how neural networks work. A neuron takes in a number of inputs and it gives out just one output, similar to a perceptron. For it to get the output, it has to calculate a value designated as *s* by multiplying each input with a different weight, which are then added together, and an additional number named the bias is then added. Mathematically, *s* is calculated as follows:

s= weight1 * input1....+ weightn * inputn + bias

In this case, our neuron will give us just *s*, but this becomes boring as the *s* will be just a linear function. However, the environment is dynamic, and we should come up with a way to model it, and our above method is very inflexible. This calls for us to add an additional step, which is known as the activation function. The activation function is the function which will take our *s* and then give the output of the neuron, known as "activation."

However, despite this, we don't have a way of telling whether our neural network is getting either closer or further from the right answer. The solution to this problem is using an activation function which is smooth, or a differentiable one. A good example of such a function is the "sigmoid function."

The fact is that the ability of a neural network to make complex decisions is determined by its internal structure or connections. This is because the output from one neuron forms the input for another neuron, which calls for a good connection between these neurons. This is the only way we can ensure that the neurons give us the correct answer.

This means that the neurons are not interconnected randomly. To ensure efficiency, the neurons are interconnected in groups of computational units. This is why the artificial neural networks are structured into layers. These layers include the following:

- The input layer.

- One or more hidden layers

- The output layer

The input layer is the one which receives the inputs and sends them to the neural network. The number of inputs to be received is usually changed by changing the number of neurons in the input layer.

The output layer will give us the output from our neural network, and the number of neurons in this layer can be altered so as to reflect the number of outputs we need from the network.

The hidden layers are added in between the input and the output layers. The number of hidden layers varies from network to network, but this is determined by the person who is creating the neural network.

Also, it is good for you to note that each layer used in a neural network has to be fully connected to the layer before it and the layer after it.

How they Work

In neural networks, machine learning takes place in the training part. Training a neural network refers to adjusting the values or the parameters, which are the weights and the bias until you get the correct answer. The process of adjusting the parameters for the neural network makes machine learning a very complicated aspect in computing. This calls for us to define a "cost function" which will help us define how wrong our neural network is after we get an output from it.

An example of this is when you have a neural network for recognizing images. If you feed a mango to it, then it tells you it sees an orange, and then the cost for your example will be too high. To ensure that the neural network gives the correct answer, thousands of examples are used for training. After getting the cost function, a "gradient descent" is performed so as to help minimize our cost function by adjusting the necessary parameters.

The gradient descent provides us with a way to determine the minimum of a function. In this case, we need to minimize the cost function. With the gradient descent, this is achieved by adjusting the values of our network in such a way that we get a lower value from the cost function than we had before.

Backpropagation

The Backpropagation algorithm works in two phases, namely the forward pass and the backward pass. During the forward pass, you just use the inputs which you are provided as well as their weights. These values are passed through all the layers, and the final output is obtained at the output layer. This value is compared to the correct answer. This will help us determine the amount of error in the output.

This gives us the error in the output layer. We calculate the error in the output layer first, because it has a direct effect on the value we obtained. However, the output layer is dependent on the hidden layer or layers. Due to this, we have to determine the error in the hidden layers by use of the gradients descent. After that, the same will also be done for the input layer. This explains why the name "backpropagation."

Chapter 5- Machine Learning in Python

In this chapter, we will be showing you how to implement machine learning in Python. We will be showing the SciPy library.

We will begin by downloading and installing the SciPy library which will help us get most libraries which are useful for machine learning in Python. We will then load the dataset. Machine learning is usually done by use of a set of statistical data. We will then build a number of machine learning models and pick the nest one. We will also build the confidence to be sure that we are getting the right accuracy.We have designed this chapter to fit you if you are a beginner to machine learning with Python. We will choose the smallest project possible with Python, which will be the classification of iris flowers.

We will be using numeric attributes, so you have to determine how to load and then handle the dataset. Since this is a classification problem, we will be using the easiest of all the supervised learning algorithms. Our project is made up of four attributes and 15 rows, and this small size will make it fit into our computer memory. Since all attributes are in similar units, no transformation or scaling is needed.

We will be doing the following:

1. Installing Python and the SciPy platform.

2. Loading our dataset.

3. Summarizing our dataset.

4. Visualizing our dataset.

5. Evaluating possible algorithms.

6. Making predictions.

Setting up Python and SciPy Library

You should install Python and the SciPy platform in your computer if you don't already have them. This is the easiest task for you to perform, so we will not get into much detail regarding it.

Installing the SciPy Libraries

Our assumption is that you have installed Python 2.7 or Python 3.5. The SciPy platform has a number of libraries, and the following libraries will be needed for this project:

- Scipy
- numpy
- matplotlib
- pandas
- sklearn

There are a number of ways that you can install these libraries. The best way is for you to choose one of the methods and then follow it so as to install all the libraries.
For the users of the Mac OS X, the macports can help you to install the Python 2.7 as well as all the necessary packages. For Linux users, package managers can help you install these. A good example of this is yum in Fedora. For Windows users, the free version of the Anaconda provides you with everything that you need, so you can take advantage of this and use it.

Start Python

Before you can begin to work on the project, it is good for you to ensure that everything was successfully installed.

Start the command line of your OS, and then launch the Python interpreter by typing the following command:

Python

It is always recommended that you run your scripts on the command line rather than writing them in text editors and big IDEs.

We want to import each of the SciPy libraries into the working environment, and then check for their versions. You can type the following on the Python interpreter terminal or just copy and paste it there:

```
# Check for Python version
import sys
print('Python: {}'.format(sys.version))
# check for scipy version
import scipy
print('scipy: {}'.format(scipy.__version__))
#check for numpy version
import numpy
print('numpy: {}'.format(numpy.__version__))
# check for matplotlib version
import matplotlib
print('matplotlib: {}'.format(matplotlib.__version__))
# check for pandas version
import pandas
print('pandas: {}'.format(pandas.__version__))
# check for scikit-learn version
import sklearn
print('sklearn: {}'.format(sklearn.__version__))
```

If the installation was successful, you will get the version for each library which you are using on your system. It will also be good for you to ensure that you are using the latest version, although these do not frequently change. In case the above code gives you an error, then don't proceed, fix everything before moving to the next step.

Data Loading

In this case, we will be using the dataset for iris flowers. It is a very common dataset which is highly used in machine learning to create the "hello world" example.

The dataset is made up of 150 observations of the iris flower. The flowers have four columns of measurements which are expressed in centimeters. The fifth column shows the flower species which has been observed. Note that the dataset has only three species of flowers.

Importing Libraries

In this project, we will be using a number of projects. Let us begin by importing them into the workspace.

```
import pandas
from pandas.tools.plotting import scatter_matrix
import matplotlib.pyplot as plt
from sklearn import model_selection
from sklearn.metrics import classification_report
from sklearn.metrics import confusion_matrix
from sklearn.metrics import accuracy_score
from sklearn.linear_model import LogisticRegression
from sklearn.tree import DecisionTreeClassifier
from sklearn.neighbors import KNeighborsClassifier

from sklearn.discriminant_analysis import LinearDiscriminantAnalysis

from sklearn.naive_bayes import GaussianNB
from sklearn.svm import SVC
```

You should have everything working correctly; otherwise, you will have to do something on the SciPy library. All the above imported libraries will work together to ensure that our machine learning example runs correctly.

Loading the Dataset

Now that we have imported all the necessary libraries, it is good for us to load our dataset into the work environment. This data will be loaded directly from the UCI Machine Learning repository.

We will use Pandas for the purpose of loading the data. This is why we imported Pandas above in the line "import pandas." The Pandas will also help us explore the data using both data visualization and descriptive statistics. During the data exploration, the name for each column will be needed, and this is why each column is being specified during the loading of the data. The following code will help us load the dataset:

url = https://archive.ics.uci.edu/ml/machine-learning-databases/iris/iris.data

names = ['sepal-length', 'sepal-width', 'petal-length', 'petal-width', 'class']

dataset = pandas.read_csv(url, names=names)

Note that we have begun by specifying the URL which we will be getting the dataset from. This is the UCI Machine Learning databases. The names of the columns have then been specified, so the dataset has a total of five columns. The dataset should then be loaded.

For those experiencing problems with the network, you can choose to download the dataset to your local machine, and then change the above URL to where you have kept it in your local machine. This should work!

Summarizing the Dataset

We now need to have a look at our data. We will look at this data in the following ways:

1. Dimensions of our dataset.

2. View of the data.

3. A statistical summary of data attributes.

4. Data broken down into class variables.

Each of the above represents a single command.

Dataset Dimensions

For us to get the dataset dimensions, we can think of it in terms of attributes (columns) and instances (rows). The "shape" property will help us get these dimensions of our dataset as shown below:

```
# shape
print(dataset.shape)
```

This will give you a total of five attributes and 150 instances.

(150, 5)

View of the Data

We need to show the first 30 rows of our data. This can be achieved by use of the "head" property as shown below:

```
# head
print(dataset.head(30))
```

You should get the first 30 rows of the dataset.

Statistical Summary

We can now go ahead and get the statistical summary for all of our attributes. Some of the summary statistics we can get from these include the mean, count, min, and the max and other percentiles.
These can be obtained by use of the "describe" property as shown below:

```
# descriptions
print(dataset.describe())
```

You will notice that all the numerical values are using the same unit of measurement, which are the centimeters, and they are all found between 0 and 8.

Class Distribution

We now need to determine the number of rows or instances which belong to same class. This can be achieved by use of the "groupby" object and then supply the mode or the property we will use to group the instances. In this case, we will be using the "class." You should have the following:

**# class distribution
print(dataset.groupby('class').size())**

You will then notice that there are three classes in the dataset, each with 50 instances.

Data Visualization

Since we are now aware of some ideas regarding our data, so it is possible for us to extend this via visualizations. We are going to discuss two types of plots:

- Univariate plots will help us to understand the attributes better.

- Multivariate plots will help us to understand relationships between various attributes better.

Univariate Plots

These are the plots for every individual variable. Since we have numeric input variables, it is possible for us to create the whisker and box plots for each. This is shown below:

 # whisker and box plots

dataset.plot(kind='box', subplots=True, layout=(2,2), sharex=False, sharey=False)

plt.show()

This will give us a good idea regarding the way the input attributes are distributed. A histogram is also good for showing us a clear picture regarding how each input variable is distributed. The plotting of the histogram should be done for each input variable:

 # histograms
dataset.hist()
plt.show()

You will realize there are two input variables having a Gaussian distribution. You should note that clearly as there are algorithms which can be used for exploitation of the assumption.

Multivariate Plots

This is the best time for us to look at how the variables interact with each other. We can first have a look at the scatterplots for all the pairs of attributes. This is good for us if we need to identify the structured relationships between the input variables as shown below:

 # scatter plot matrix
scatter_matrix(dataset)
plt.show()

It is good for you to note that some attributes have been grouped diagonally. This is a clear indication of a high correlation, as well as a predictable relationship.

Evaluating Algorithms

We are going to create some models for our dataset, and then estimate the accuracy based on unseen data. We are going to follow the steps given below:

1. Separate the validation dataset.

2. Setup a test-harness to use a 10-fold cross validation.

3. Build five models for predicting the species from the flower measurements.

4. Choose the best model.

Creating a Validation Data Set

We should know whether we created a good model or not. Statistical methods will then be used later for estimation of the accuracy of our models which will be created on the unseen data.

The loaded dataset will be divided into two, whereby 80% of it will be used for training the models and 20% of it will be used for holding back as the validation dataset. This is shown below:

 # Split-out the validation dataset
array = dataset.values
X = array[:,0:4]
Y = array[:,4]
validation_size = 0.20
seed = 7

X_train, X_validation, Y_train, Y_validation =
model_selection.train_test_split(X, Y,
test_size=validation_size, random_state=seed)

After the above data, the X_train and Y_train will have the training data for preparation of the models as well as the X_validation and Y_validation which can be used later.

Test Harness

The accuracy of the models will be estimated by use of a 10-fold cross validation. With this, the dataset will be split into 10 parts, whereby the training will be done on 9 of them and testing on 1 of them. This will then be repeated for all the combinations of the train-test splits.

 # Test options and the evaluation metric
seed = 7
scoring = 'accuracy'

The metric "accuracy" has been used for evaluating the models. It refers to the ratio of the number of instances which have been predicted accurately, which should then be divided by the number of instances which are contained in the dataset, and then multiplied by 100 to convert into a percentage. The "scoring" variable will be used when running the build and when evaluating each model next.

Building Models

Currently, we are not aware of the best algorithms to use for the problem, or the best configurations which should be done. The following are the algorithms which we need to evaluate:

- Logistic Regression (LR)

- K-Nearest Neighbors (KNN).

- Linear Discriminant Analysis (LDA)

- Classification and Regression Trees (CART).

- Support Vector Machines (SVM).

- Gaussian Naive Bayes (NB).

You are aware of some of the algorithms, as we had discussed them earlier on. You can as well go back and read about them if you are not very familiar with how they operate.

The LR and LDA are non-linear algorithms, while the KNN, CART, SVM, and NB are linear algorithms. The random number seed should first be reset before every run as a way of ensuring that evaluation of every algorithm is done by the use of similar data splits. This will also ensure that we have results which can be directly compared.

We can then build and evaluate our five models:

```
# Spot Checking the Algorithms
models = []
models.append(('LR', LogisticRegression()))
models.append(('LDA',
LinearDiscriminantAnalysis()))
```

```
models.append(('KNN', KNeighborsClassifier()))
models.append(('CART', DecisionTreeClassifier()))
models.append(('NB', GaussianNB()))
models.append(('SVM', SVC()))
# evaluating every model
results = []
names = []
for name, model in models:
    kfold = model_selection.KFold(n_splits=10,
random_state=seed)

    cv_results =
model_selection.cross_val_score(model, X_train,
Y_train, cv=kfold, scoring=scoring)

    results.append(cv_results)
    names.append(name)
    msg = "%s: %f (%f)" % (name,
cv_results.mean(), cv_results.std())

    print(msg)
```

Choosing the Best Model

At this point, we have 6 models as well as the estimation of accuracy for each. Our goal is to perform a comparison on all the available models and then pick the best one based on accuracy. The example should give you the result shown below:

```
LR: 0.966667 (0.040825)
LDA: 0.975000 (0.038188)
KNN: 0.983333 (0.033333)
CART: 0.975000 (0.038188)
NB: 0.975000 (0.053359)
SVM: 0.981667 (0.025000)
```

From the above results, KNN seems to be the one with the highest estimation accuracy. It is now possible for us to create a plot from the results given above and then perform a comparison on the mean accuracy and the spread of the results.

```
# Compare the Algorithms
fig = plt.figure()
fig.suptitle('Comparison of Algorthms')
ax = fig.add_subplot(111)
plt.boxplot(results)
ax.set_xticklabels(names)
plt.show()
```

Making the Predictions

The model for KNN was the most accurate one among the tested ones. It is now time for us to get an idea regarding the accuracy of our model on the validation set. It is advisable that you keep a validation set as a slip may occur during the training process.

The KNN model can now be run directly on our validation set, and then the results from this should be summarized as a final accuracy score, classification report, and a confusion matrix.

```
# Make some predictions on the validation dataset
knn = KNeighborsClassifier()
knn.fit(X_train, Y_train)
predictions = knn.predict(X_validation)
print(accuracy_score(Y_validation, predictions))
print(confusion_matrix(Y_validation, predictions))
print(classification_report(Y_validation,
predictions))
```

The accuracy will be 0.9 or 90%. A confusion matrix is good for giving an indication of the three errors made. The classification report provides us with a breakdown of every class by precision, recall, the f1-score, and the support so as to give excellent results.

These are all shown below:

0.9

```
[[ 7  0  0]
 [ 0 11  1]
 [ 0  2  9]]
```

	precision	recall	f1-score	support
Iris-setosa	1.00	1.00	1.00	7
Iris-versicolor	0.85	0.92	0.88	12
Iris-virginica	0.90	0.82	0.86	11
avg / total	0.90	0.90	0.90	30

That is it! You now have a fully working machine learning project. Congratulations!

Chapter 6- Clustering

Remember our Iris dataset. Suppose we are aware that we have three types of Iris, but we don't a taxonomist for labeling them. A clustering task can help us in this case place the observations into groups which are well organized, referred to as "clusters."

K-Means Clustering

There are many criteria and algorithms for clustering, but K-Means is the simplest clustering algorithm.

In Python, this can be used as shown below:

```
from sklearn import cluster, datasets
iris = datasets.load_iris()
X_iris = iris.data
y_iris = iris.target

k_means = cluster.KMeans(n_clusters=3)
k_means.fit(X_iris)

print(k_means.labels_[::10])

print(y_iris[::10])
```

Note the it may be difficult for you to choose the right number for the clusters, but in this case, we have chosen three. The K-Means algorithm is also highly sensitive when it comes to initialization, and it may fall into the local minima, but scikit-learn has a number of tricks which helps to mitigate this risk.

Vector Quantization

In K-Means, clustering can be seen as a way of choosing small exemplars from a population so as to compress your information. This is sometimes known as vector quantization. We can employ this mechanism when we need to posterize an image as shown below:

```
import scipy as sp
try:
  face = sp.face(gray=True)
except AttributeError:
  from scipy import misc
  face = misc.face(gray=True)
X = face.reshape((-1, 1)) # We are need of an (n_sample, n_feature) array

k_means = cluster.KMeans(n_clusters=5, n_init=1)
k_means.fit(X)

values = k_means.cluster_centers_.squeeze()
labels = k_means.labels_
face_compressed = np.choose(labels, values)
face_compressed.shape = face.shape
```

In this case, we are creating five clusters. In hierarchical clustering, the clusters are created and organized into a cluster. There are two best approaches in this:

1. Agglomerative- this is for bottom-up approaches. Every observation will begin in its own cluster, and the merging of the clusters is done iteratively so as to minimize the linkage criterion. This is the best approach if the cluster is composed of a few number of observations.

2. Divisive- this is for top-down approaches. All the observations will begin in one cluster, which is split iteratively while moving downwards the hierarchy. This approach becomes slow when we are estimating large numbers of clusters. The reason behind this is that all the clusters usually start as a single cluster, which is then split iteratively.

Connectivity-constrained clustering

In the case of agglomerative clustering, you can give a connectivity graph so as to give the samples which you are able to classify. In scikit, graphs are represented by an adjacency matrix. A sparse matrix is widely used for this. This becomes very useful when you need to retrieve the connected regions during the clustering of an image. This is demonstrated below:

```python
import matplotlib.pyplot as plt
from sklearn.feature_extraction.image import
grid_to_graph

from sklearn.cluster import AgglomerativeClustering
from sklearn.utils.testing import SkipTest
from sklearn.utils.fixes import sp_version

if sp_version < (0, 12):

    raise SkipTest("Skipping  since SciPy versions
being used is earlier than 0.12.0 and "

            "and it does not have the scipy.misc.face()
image.")

  # Generate the data
try:
    face = sp.face(gray=True)
except AttributeError:
    # Newer versions of the scipy have face in the misc
    from scipy import misc
    face = misc.face(gray=True)

# Reduce its size to 10% of original size so as to speed
up processing

face = sp.misc.imresize(face, 0.10) / 255.
```

Note that we expect that you should be using a SciPy version which is not less than 0.12.0. If this is not the case in your system, then the interpreter will skip, as you will not have the scipy.misc.face() image. We have also reduced the size of the image up to 10%. This is because with a small-sized image, we will get a high speed of processing, and this is what we need. This is what we have done in this case, thus, you will get a high speed of processing.

Feature agglomeration

Sparsity is a good way of mitigating a curse of dimensionality, that is, when we have an insufficient number of observations compared to the number of features. You can also choose to merge together features which are similar, which is known as feature agglomeration. We can easily achieve this feature by clustering the data which is being transported. This is shown below:

```
digits = datasets.load_digits()
images = digits.images
X = np.reshape(images, (len(images), -1))
connectivity = grid_to_graph(*images[0].shape)

agglo =
cluster.FeatureAgglomeration(connectivity=connecti
vity,
                    n_clusters=32)
agglo.fit(X)

X_reduced = agglo.transform(X)

X_approx = agglo.inverse_transform(X_reduced)
images_approx = np.reshape(X_approx,
images.shape)
```

In the case of some estimators, they will expose a "transform" method as a way of reducing the dimensionality of a dataset.

Conclusion

We have come to the end of this book. Machine learning is of great importance in today's world. It is with machine learning that we are able to create a system which acts as humans. A good example of such a system is a driverless car. Such a car can be made in such a way that it obeys traffic as if it is being driven by a human being. We can also create systems capable of recognizing human speech. Machine learning systems usually learn from experience and through interaction with other systems, and this makes them show an improvement over time. Algorithms for machine learning have been developed to help in training machine learning systems. These algorithms can be classified into either supervised or unsupervised algorithms. In supervised learning algorithms, training data is provided, but in the unsupervised learning algorithms, no training data is provided.

www.ingramcontent.com/pod-product-compliance
Lightning Source LLC
Chambersburg PA
CBHW051259170526
45165CB00004B/1773